REALITY

**Distinguishing Real and Imaginary States of Affairs for
Recognizing Generally Accepted Falsehoods and Misconceptions**

an

Un-Maa Saa Publication

Copyright © 2015 Un-Maa Saa

1

Table of Contents

Table of Contents
(Page 2)

Introduction
(Page 5)

Definitions
(Page 10)

Section One: Reality
1: Possibility
(Page 12)

2: Social Subversion
(Page 15)

3: Morality
(Page 17)

4: Spirit
(Page 23)

5: Emotion
(Page 25)

6: Publicly Unobservable Sentience
(Page 26)

7: Unobstructed Desire
(Page 30)

8: Infinite Continuum
(Page 31)

9: Sentience and the Sensational World
(Page 34)

10: Omni-Personhood
(Page 36)

11: The Gods
(Page 39)

12: Dishonesty
(Page 45)

Section Two: R-Theory
1: Gravitation
(Page 62)

2: Time
(Page 64)

3: Space
(Page 65)

4: Motion
(Page 67)

5: Gravitational Singularities
(Page 69)

6: Cosmological Constant
(Page 71)

7: Hyperspace
(Page 73)

Section Three: Standard Radix
(Page 75)

Quantum Base Units
(Page 77)

Mejasinu Numerals
(Page 77)

Introduction

Genesis 1:6-8
And Elohiym said, "Let there be a **pound-out** between the waters to separate water from water." So Elohiym made the **pound-out** and separated the water under the **pound-out** from the water above it. And it was so. Elohiym called the **pound-out** "sky." And there was evening, and there was morning—the second day.

Genesis 1:14-19
And Elohiym said, "Let there be lights in the **pound-out** of the sky to separate the day from the night, and let them serve as signs to mark sacred times, and days and years, and let them be lights in the **pound-out** of the sky to give light on the earth." And it was so. Elohiym made two great lights—the greater light to govern the day and the lesser light to govern the night. He also made the stars. Elohiym set them in the **pound-out** of the sky to give light on the earth, to govern the day and the night, and to separate light from darkness. And Elohiym saw that it was good. And there was evening, and there was morning—the fourth day.

According to the author of Genesis 1, the sky had been created as a **pound-out** with water both below and above the **pound-out** in which the sun, the moon, and the stars had all been embedded. The word traditionally transliterated into the English language as '**firmament**' or '**expanse**' is the word '**raqiya**' which literally means '**pound-out**' in reference to a malleable metallic substance to have been hammered out into a sheet spread over the earth as a dome.

FACT: the sky is not a malleable metallic substance to have been pounded out over the earth.

FACT: the celestial bodies have never been embedded lights in a

5

pounded out substance.

FACT: no division of terrestrial water has ever occupied a space above the earth that is also beyond the space of the sun, the moon, and the stars as the limits of the earth's atmospheric water do not extend the distance of the moon, the sun, or the stars.

The cosmological perspective apparent in the Genesis 1 creation account has been out of date since at least the 4th century B.C.E. when it had become obvious that the concept of a solid domed sky did not conform with the observations of celestial movements. Even hundreds of years before this time, knowledge of solar eclipses involving the passage of the moon between the sun and the earth had clearly revealed that the two most visible celestial bodies had never occupied a single distance from the earth as required for the single layer of a domed pound-out. **Clearly**, especially in a world of air travel and artificial orbiting satellites, any modern assertions of terrestrial water being beyond the space of the celestial bodies or of the sky being a pounded out dome above the earth are either plainly dishonest or clinically delusional. **Clearly**, the reality is that no proposed creator deity such as Elohiym had ever accomplished the non-existing celestial states of affairs ascribed to him in Genesis 1. **Clearly**, any physically undetectable literary character portrayed as <u>having validated non-existing celestial states of affairs</u> by the character's own quoted commands is either a character of fiction produced by an uninformed imagination ignorant of celestial reality or the product of a conscious endeavor of deceit. **Clearly**, fictional accounts written by demonstrably uninformed ancients do not provide a reliable source of knowledge about the world and cannot be a viable alternative to honest observations and reason. Impossible imaginary perspectives have no validity in the light of genuine experiences with reality and imaginary storybook characters possess no ability to accomplish anything in the real world.

That the creator-god of the Genesis 1 creation account is an imaginary storybook character is evident not only from the character's direct association with acts of fiction in imaginary stories but is also evident upon distinguishing the conditions of real and imaginary states of affairs. As with the perspective of the biblical god, gravitation is the most powerful force within the universe and is known to be responsible for the existence of all the celestial bodies in the universe. Like the biblical god, gravitation is irresistibly felt by all matter and is omnipresent though it cannot be seen. Gravitation cannot be heard. Gravitation cannot be touched. Gravitation cannot be smelt or tasted. No instrument can detect a fundamental unit of gravitation. Yet, unlike the biblical god, at any given moment, at any given place on earth, the ever present reality of gravitation can be publicly demonstrated and even observed in the universe to reveal the existence of dark matter that is otherwise intangible to detection. Questions exist as to the nature of gravitation but there is never any question or doubt as to the existence of gravitation due to its ongoing demonstrable presence. That's the difference between a real and an imaginary subject. No matter how intangible, real subjects have a publicly discernible interaction with the observable world that can therefore be publicly shown or demonstrated and therefore be publicly known while imaginary subjects, be they widely accepted characters from a book of stories or otherwise, never have a demonstrable presence in the publicly observable world. Imaginary subjects never exist outside of the worlds of the stories they're presented in or outside of the minds of the individuals to imagine them. This differs from hypothetical subjects that are imaginary yet possess qualities that may allow them to be tested in the observable world and are therefore possible subjects until shown to be invalid.

To explain the existence of dark matter, a phenomenon known to exist only from the deduction of observed gravitational effects, an imaginary subject that can be called a neutrono can be imagined to be composed of positive/negative pairs of another imaginary subject that can be called leptuarks that have individual masses approximately 1/3 of a down quark and a 3/3 charge. As with quarks, leptuarks would be subject to the same quantum forces

7

along with gravitation and exist in three generations. However, unlike purely imaginary subjects, leptuarks would possess properties that are actually known to exist and can therefore be evaluated with respect to the observable world to determine the status of their validity making them hypothetical. Imaginary subjects without any known properties that can either be directly observed or deduced from consistently observed results are non-hypothetical subjects beyond the realm of actual knowledge and known possibility (as knowing what is possible is dependent upon knowing the conditions that allow a subject to be possible). Any subject claimed to be possible without there being any knowledge of any known conditions that allow the subject's possibility is a subject that is literally not known to be in any way possible and is therefore an impossible subject (as any knowledge pertaining to the subject violates conditions known to be real). Any impossible subject claimed to influence the publicly observable world with no consistent means of somehow publicly observing the interaction of the subject is a subject that is beyond any possible public knowledge of it and is therefore a subject confined to private psychological experience that is otherwise known to be an experience of the imagination. There is simply no such thing as having knowledge of a subject that is external to private psychological experience yet cannot be externally detected or deduced since knowledge of any subject external to private psychological experience requires that the subject be detectable outside of private psychological experience in order to be knowable. In short, subjects known to be real have a publicly discernible interaction with the observable world while subjects known to be purely imaginary are neither publicly discernible nor known to be in any way possible. In turn, a **Realist** is one who affirms the reality of real states of affairs while a **Denialist** is one who selectively denies or ignores the reality of particular real states of affairs in preference for particular imaginary states of affairs. As such, the notion of a realist differs from the notion of an atheist in that a realist is defined by a natural acceptance of reality while an atheist is specifically defined by a lack of belief in a particular imaginary notion. Being an atheist does not preclude holding beliefs in imaginary notions such as karma and

8

samsara while being a realist does. A realist lacks belief in all imaginary notions, not just one. Since individuals do not actually center their perspectives around lacking a belief in any particular notion, 'atheist' is merely one of multiple descriptions that can apply to an individual whose functionality in life is by no means dependent upon that description. Such individuals are only perceived to be functionally centered upon that description from the restricted perspectives of those for whom a particular imaginary notion is held to be of great significance.

That the biblical god is indeed an imaginary notion is by no means an exaggerated characterization or mere opinion of perspective but is a literal matter of fact, honestly undeniable by the distinction of real and imaginary, being a notion not in any way experienced outside of the imagination. That the biblical god had been presented as having engaged in impossible acts of creation that the author of the Genesis 1 creation account had simply imagined yet had nevertheless presented as being actual events shows an intentional endeavor of deceit in operation at the very beginning of the Bible that goes far beyond any mere pious act of ignorance. The distinguished conditions pertaining to real and imaginary states of affairs not only allows recognition of imaginary subjects but also allows recognition of blatantly dishonest intents. To aid in the exposure of dishonest intents commonly employed by either dishonest or delusional individuals, the following pages in Section One of Reality presents groups of simple individual truth statements that have wide application to generally promoted perspectives commonly asserted to be either valid or plausible. In short, Reality is the realist's handbook of simple recognizable truths for exposing dishonest intents and delusional perspectives.

Definitions

Real – *that which is either publicly perceivable or deducible from observably consistent phenomena.*

Non-Existent – *that which is neither publicly perceivable nor deducible from observably consistent phenomena nor known to be in any way possible.*

Imaginary – *any subject of a mentally entertained state of affairs that is neither publicly perceivable nor directly deducible from observably consistent phenomena nor known to be in any way possible.*

Delusional – *a psychological condition pertaining to an inability to distinguish an imagined state of affairs from a real state of affairs.*

Possible – *any proposed state of affairs known to be consistent with the properties of observably consistent phenomena.*

Impossible – *any proposed state of affairs involving mutually exclusive conditions.*

Truth – *an asserted state of affairs known to be real.*

Falsehood – *an asserted state of affairs known not to be real.*

Honest – *promoting a presented state of affairs that the promoter believes to be true.*

Dishonest – *promoting a presented state of affairs that the promoter consciously does not know to be true.*

Morality - *an individual's personal commitment to preventing unnecessary misfortune.*

Amorality - *an individual's willingness to allow unnecessary misfortune.*

Immorality - *an individual's willingness to cause unnecessary misfortune.*

Emotions – *psychological reactions to personal experiences of gratification and/or discontent.*

Realist – *a person who affirms the reality of real states of affairs, comprehending the world through natural observation and reason.*

Denialist - *a person who selectively denies or ignores the reality of particular real states of affairs in preference for particular imaginary states of affairs.*

Section One:

Reality

Possibility

1:1

Everything publicly known to be real is known through publicly perceptible phenomena.

1:2

Publicly perceptible phenomena are the only means by which anything is publicly known to be real.

1:3

Nothing is publicly known to be real without the presence of publicly perceptible phenomena.

1:4

Nothing is publicly known to be possible without the presence of publicly perceptible phenomena possessing properties revealing the possibilities.

1:5

Publicly perceptible phenomena are the sole fundamental basis upon which anything is publicly known to be possible.

1:6

Anything not publicly known to be possible is by consequence publicly known to be impossible as all public knowledge pertaining to that subject results in impossibility.

1:7

Only in revealing publicly perceptible phenomena that is contrary to that which is publicly known to be impossible can that which is publicly known to be impossible be shown to be possible.

1:8

The subjects of concepts that are not revealed through publicly perceptible phenomena are imaginary subjects as a conceived subject is an imagined subject and a subject that is never publicly encountered apart from the imagination is an imaginary subject.

1:9

Subjects are neither real nor possible simply because they are imagined to be.

1:10

Imaginary subjects not known to be possible cannot be responsible for the existence of observable subjects.

1:11

Subjects that are not in any way perceptible and are not in any way deducible are subjects that are not in any way knowable.

1:12

Subjects that are publicly known not to be possible and publicly known not to have any public demonstration are subjects that are publicly known not to be real.

1:13

An imagined notion naturally elicits no consideration for its existence unless its possibility can be demonstrated resulting in the natural fact there is never any need to disprove anything not known to exist.

1:14

An honest assertion for the possibility of a proposed state of affairs is predicated upon knowledge of the conditions that allow the state of affairs to be possible.

1:15

An assertion for the possibility of a proposed state of affairs without knowledge of the conditions that would allow the state of affairs to be possible is an assertion of falsehood that affirms knowledge of conditions that in fact are not known.

1:16

Any assertion for the existence of a state of affairs without an observation of the state of affairs or knowledge of any possibility for the state of affairs is a falsely made assertion.

Social Subversion

2:1

To imagine a state of affairs and then accept the validity of that state of affairs against observation and deductive reasoning is a psychologically unhealthy approach to thinking known as delusion that puts individuals at odds with reality and at risk to any detrimental situations to result from a delusional approach to life.

2:2

The promotion of a delusional approach to life is both socially subversive and immoral and should be exposed if not prohibited in accordance with public safety to protect innocent individuals from the possibilities of psychological and physical harm.

2:3

Individuals who are under the influence of a delusion should be eligible for cognitive therapy to assist the individuals in confronting the emotional issues to have impaired their psychological health.

2:4

Individuals who facilitate the maintenance of delusional perspectives in others in disregard to observation and reason are engaged in dishonest endeavors of social subversion to prohibit the freedom of thought capable of maintaining social independence from deception-based social authorities.

2:5

To consciously serve in directing individuals away from available opportunities for knowing reality by reinforcing their ignorance of reality is to consciously serve in an immoral endeavor of social subversion.

2:6

Any endeavor to maintain a state of ignorance in individuals in order to keep them subjected to deception by a perceived authority is an immoral endeavor.

Morality

3:1

Prehistoric misidentifications of the wind with breathing and death with sleeping do not establish reality for beliefs in an anthropomorphism of the sky, postmortem disembodied habitation of the air, or an eternally blissful state of dreaming.

3:2

Adherers of socially embraced delusional perspectives that are claimed to be moral are seen to be no less prone to immoral behavior than adheres of reality.

3:3

The embrace of reality cannot ensure a person with an eternally blissful afterlife.

3:4

The embrace of reality cannot ensure a person with a good life.

3:5

The embrace of reality can only ensure a person with a degree of clarity in their interaction with the world for utilization in their own best interests.

3:6

Consciously misrepresenting imaginary states of affairs to

individuals as being real states of affairs reveals a willingness to allow unnecessary misfortune for individuals through a conscious act of deception and therefore a willingness to facilitate conditions of unnecessary misfortune for those individuals.

3:7

To deny a harmless person knowledge of reality is to undermine their ability to independently act in their own best interests for enjoying the only life they have to live as deceiving individuals imparts individuals with false information that prohibits the individuals from making accurate assessments of their circumstances thereby facilitating the conditions of making misinformed decisions to result in unnecessary misfortune.

3:8

To convince a harmless person that an eternally blissful life exists beyond their mortal existence is to facilitate conditions by which that person may allow unnecessary suffering throughout their life or even allow premature death in expectation of an imagined life to follow a consequentially miserable existence that need not have been so miserable.

3:9

To consciously facilitate a consequentially miserable existence for individuals that need not have been so miserable due to a social endeavor of dishonesty is to facilitate an immoral endeavor at odds with the advocation of genuine morality.

3:10

Morality is an individual's personal commitment to preventing unnecessary misfortune.

3:11

A personal commitment to preventing unnecessary misfortune does not result from the facilitation of psychological coercion through rewards and punishments as personal commitments are internally initiated and therefore are not the result of external coercion.

3:12

Facilitating acts of morality in individuals through the coercion of rewards and punishments, real or imaginary, facilitates conditions in which individuals may be inclined to mainly act amorally without a reward to be received or a punishment to be avoided.

3:13

Promotion of the perspective that a specific unsubstantiated belief must be accepted apart from any degree of personal morality in order for an individual to be personally justified is the promotion of a perspective that rates the importance of a belief above the importance of morality.

3:14

Promotion of the perspective that morality is not internally initiated and can only be imposed from without facilitates perspectives in individuals that moral compliance can only be achieved through a state of coercion that is normative for interaction.

3:15

Promotion of the perspective that true morality and knowledge of the world can only be identified with a specific unquestionable ideology open to the employment of coercion for compliance facilitates a self-perspective of moral superiority and social entitlement with unquestionable justification for the use of coercion in bringing about ideological compliance.

3:16

Obedience to a perceived power simply because it is perceived to be powerful regardless of the perceived power's actual morality is not in any way morality as an internally initiated commitment to preventing unnecessary misfortune is not an issue dependent upon reaction to an external power and the disregard of actual morality is obviously not morality.

3:17

Convincing individuals to adhere to an imaginary subject requiring obedience to immoral directives facilitates the acceptance of such sanctioned immoral directives as being acceptable moral behavior.

3:18

Worship of an imaginary subject conceived of as being a supernatural entity that either commands or engages in the immoral acts of genocide, infanticide, murder, enslavement, gender subjugation, coercion, arbitrary social impositions and restrictions, land theft, lying, and condemnation for non-acceptance of unsubstantiated claims, is the worship of a subject identifiable with the definition of a demon.

3:19

Facilitating sociological conditions known to be conducive to immoral behavior is an immoral endeavor.

3:20

A genuine advocation of morality does not involve a promise of postmortem rewards or postmortem threats of severe consequences or postmortem karmic rebirths, all assured from social endeavors of dishonesty.

3:21

A genuine advocation of morality does not appeal to promises of personal rewards and punishments as the motivation for behavior that is at best amoral.

3:22

A genuine advocation of morality does not require the acceptance of any beliefs for moral justification or value any belief above morality.

3:23

A genuine advocation of morality does not provide sanction for extremist acts of immorality against individuals who don't embrace a socially delusional perspective.

3:24

A genuine advocation of morality does not promote the honor or worship of individuals who command or engage in genocide, infanticide, murder, enslavement, gender subjugation, coercion, arbitrary social impositions and restrictions, land theft, lying, and condemnation for non-acceptance of unsubstantiated claims.

3:25

A genuine advocation of morality, as a social endeavor of preventing unnecessary misfortune, is a social endeavor of honesty inclusive of preventing the unnecessary misfortunes of dishonesty while advocating morally applied knowledge of reality.

3:26

Any worldviews to facilitate selective departures from reality and selective departures from morality are worldviews to facilitate unnecessary misfortune that any moral individual would be personally committed to preventing.

Spirit

4:1

A localized activity is constituted upon energy or mass undergoing a change in space of limited range.

4:2

A localized activity does not exist independent of energy or mass undergoing a change in space of limited range.

4:3

Mass (m) itself is an activity (E/c^2) localized to a limited range of space through the Higgs field.

4:4

The configuration of a localized activity can be perpetuated in masses other than the original mass of the localized activity (e.g. genetic replication).

4:5

The configuration of a localized activity cannot be perpetuated in a mass that is intangible to the mass of the localized activity.

4:6

A subject that is tangible to the observable world cannot be in possession of a subject that is intangible to the observable world.

4:7

A spirit/soul that is intangible to the observable world cannot be in the possession of a subject that is tangible to the observable world.

4:8

A spirit/soul identified as being a localized activity of the observable world cannot be perpetuated in a subject that is intangible to the observable world.

Emotion

5:1

Sentience is existence.

5:2

The only subject that an individual can know yet be unable to prove is their psychological state as revealed by the fact that an individual knows their thoughts and emotional state and can express and report their thoughts and emotional state but cannot prove their thoughts and emotional state to anyone who cannot directly perceive an individual's thoughts and emotional state and therefore cannot know if an individual is either acting or lying about their thoughts and emotional state.

5:3

Public knowledge of the objective existence of emotion that is otherwise only privately known to every individual is publicly known from the consistent public observation of emotion expressed in individuals and is therefore publicly known to be an existence confined to the psychological experience of individuals.

5:4

Emotion is an observable physical manifestation of an individual's personhood expressed through independent reactions from which its independent presence can be deduced.

Publicly Unobservable Sentience

6:1

Every action experienced by a subject produces an equal and opposite reaction to the subject being experienced.

6:2

A subject cannot experience an action from a subject that does not experience an equal and opposite reaction.

6:3

A subject cannot be tangible to the experience of an action without being tangible to the cause of the action.

6:4

To sense or influence a subject is to interact with a subject through means of shared physical properties.

6:5

A sentience cannot see or be seen in the observable world unless the sentience shares physical properties in the observable world capable of receiving or reflecting electromagnetic waves.

6:6

A sentience cannot hear or be heard in the observable world

unless the sentience shares physical properties in the observable world capable of detecting or producing pressure waves.

6:7

A sentience cannot directly communicate verifiable information of the observable world to a sentience of the observable world unless the sentience shares physical properties in the observable world capable of receiving and producing sensations.

6:8

A sentience cannot affect the motion of matter in the observable world unless the sentience shares physical properties in the observable world capable of reacting to either the mass or the charged forces that compose matter.

6:9

A sentience cannot maintain a consistent location in the observable world without the application of great force for controlled motion unless the sentience shares physical properties in the observable world that subject it to the effects of gravitation.

6:10

Anything becoming tangible to the observable world from an intangible source automatically loses its connection with the intangible source upon becoming tangible.

6:11

Only an ongoing gravitational event is theoretically possible for being sustainable in the observable world from an intangible hyperspace source thereby making the gravitational event publicly observable and its hyperspace source publicly deducible.

6:12

The fact that an ongoing gravitational event originating from a hyperspace source would have to undergo an application of great force for controlled motion to maintain an observable location in the observable world would automatically imply that the intangible hyperspace source is under the direct control of a sentience.

6:13

The objective perception or presence of a sentience requires such a sentience to possess physical properties that make it tangible to the observable world at the spatial point of its interaction with the observable world thereby making it a publicly observable sentience during its interaction with the observable world.

6:14

An unobservable observer cannot exist as the physical properties to allow for observation render the observer observable to the interactions of those physical properties.

6:15

An observer cannot be tangible to the interactions of physical properties that allow the observer to engage in observation and at the same time be intangible to the very same interactions of those physical properties to thereby remain unobservable.

6:16

Everything publicly known can be shown.

6:17

Anything that cannot be shown cannot be publicly known.

6:18

Sentience can only be inferred from observable independent reactions.

6:19

A sentience that can only be known to an individual's thoughts and feelings, unable to be publicly observed to independently produce those specific thoughts and feelings in any other individuals for corroboration, is an imaginary sentience.

6:20

The notion of a publicly unobservable sentience capable of interacting with the observable world yet producing no observable independent reactions to the observable world from which its independent presence could be deduced is the notion of a sentience confined to only being known psychologically thereby making it an imaginary sentience.

6:21

A publicly unobservable, self-revealing, reactionary sentience, is an imaginary sentience that cannot independently exist outside of an individual's mind by virtue of the fact that being publicly non-experienced yet self-revealing to the individual mind it is revealed to renders any experienced existence of the sentience dependent upon the existence of that individual mind.

Unobstructed Desire

7:1

An unobstructed desire is always fulfilled when the possessor of the unobstructed desire has the ability to fulfill the unobstructed desire.

7:2

An assertion that a sentience has the unobstructed desire to be known to everyone accompanied by the assertion that the sentience has the ability to make itself known to everyone followed by the observation that the sentience does not make itself known to anyone means that the assertion that the sentience has the unobstructed desire to be known to everyone is false, or the assertion that the sentience has the ability to make itself known to everyone is false, or the premise that such a sentience exists is false.

Infinite Continuum

8:1

The observable world reveals the consistent decay of finite composite things into a static equilibrium in a process called entropy.

8:2

The consistent decay of finite composite things into a static equilibrium is the diffusion of the finite amount of energy upon which the finite composite things are established.

8:3

The diffusion of the finite amount of energy upon which the finite composite things are established imply a time of origin of the diffusion of energy and therefore a time of origin of the establishment of the finite composite things.

8:4

Finite composite things derive from either greater energy or from nothing.

8:5

To state that a thing can derive from nothing is to imply that the existence of a thing is of the existence of no thing.

8:6

To imply that the existence of a thing is of the existence of no thing is to imply that a thing can be of that of which it is not.

8:7

To imply that a thing can be of that of which it is not is a direct contradiction.

8:8

A thing (the presence of a thing) and no thing (the absence of a thing) are opposite and mutually exclusive conditions which imply that a thing can only be of a thing and that no thing can only derive no thing.

8:9

For any finite composite thing to exist there must always have been some greater energy to have derived it.

8:10

For some greater energy to have always existed in a non-entropic state, an infinite amount of energy must always have existed to have prohibited static equilibrium.

8:11

The existence of finite composite things cannot be derived from nothing but must be derived from greater energy that is either of, or the result of, an infinite continuum.

8:12

The theoretical deduction of an infinite multiverse corroborates the necessary conclusion of the finite observable world being derivative of an infinite continuum.

8:13

The observable world, being derivative of an infinite continuum that is by necessity eternal, is the product of an eternally natural process that precludes any notion of an intentional act of sentient origin.

Sentience and the Sensational World

9:1

Sensation is the interaction of the sensational world with the constituents of sentience.

9:2

Sentience is the organization of sensation predicated upon the intercommunication of a neural network that is tangible to the sensations being received.

9:3

The constituents of sentience require tangibility to the physical properties of sensation.

9:4

The sensational world produces sensations that are organized into sentience.

9:5

The constituents of sentience are properties of the sensational world.

9:6

Sentience is the derivative of the sensational world.

9:7

Sentience, being the organization of sensation that is produced by the sensational world, cannot exist independent of the sensational world.

9:8

No sentience can precede the existence of the sensational world as there can be no sentience without the sensations of the sensational world.

9:9

The notion of a sentient originator of the sensational world is not the notion of a possible sentience as a sentience cannot be sentient without the sensations of the sensational world to be sentient of.

Omni-Personhood

10:1

Contemplation requires a degree of ignorance of a subject being contemplated as contemplation is a process of realizing aspects of a subject of which the contemplator may previously have been unaware.

10:2

Intention requires motivation to change an insufficiently satisfactory state of affairs for which the intender has lacked the ability to preempt thereby making the intender subject to an insufficiently satisfactory state of affairs motivating the intention to change the insufficiently satisfactory state of affairs.

10:3

Any states of affairs to exist contrary to an individual's satisfaction exist beyond that individual's power of prevention.

10:4

The ignorance required for contemplation and the lack of power required for intention preclude a contemplative and intentional sentience being either omniscient or omnipotent.

10:5

A subject cannot simultaneously be all-knowing and not all-knowing.

10:6

A subject cannot simultaneously be all-powerful and not all-powerful.

10:7

An omniscient and perpetually self-sufficient omnipotent subject, naturally incapable of possessing the limitations and motivation required for contemplation and intention, is naturally devoid of the mortal responses to constitute the attributes of personhood.

10:8

The notion of an omniscient, omnipotent, and by consequence, omnipresent, person, possessing the mortal attributes of personhood, is not the notion of a possible person as it is a notion involving conditions of mutual exclusivity in which the conditions for being an omni-subject precludes the conditions for being a responsive sentience.

10:9

The notion that an omniscient, omnipotent, subject that orchestrates everything in accordance with its omniscience that can be appealed to for granting requests that aren't already in accordance with its omniscience involves the impossibility of an omniscient subject acting contrary to its own state of omniscience thereby nullifying its condition of orchestrating everything in accordance with its omniscience.

10:10

Any states of affairs to be requested of an omniscient, omnipotent, subject conceived of as orchestrating everything in accordance with its omniscience are states of affairs that are either going to happen whether or not the requests are made or are states of affairs that are not going to happen whether or not the requests are made.

The Gods

11:1

An unobserved, contemplative, omniscient, omnipotent, intentional, sentient originator of the sensational world is an impossible, imaginary, concept that is ever non-existent to natural public experience thereby relegating the concept beyond any rational reason for serious consideration.

11:2

Contemplative, intentional, sentient beings are only publicly known to be finite biological beings originating from the sensational world derived from the physical properties of the sensational world.

11:3

Anything derived from or sharing in any physical property of the sensational world is subject to that physical property for its existence and therefore cannot possess an inherent universal control over a physical property to which it is already inherently subject.

11:4

Any asserted sentience claimed to have an inherent universal control over a physical property of the sensational world cannot have an existence subject to that physical property.

11:5

Finite beings with an inherent control over a physical property of the sensational world do not exist as finite beings cannot have inherent control over physical properties to which they themselves are naturally subject.

11:6

An infinite, interventional, sentience with an inherent control over the physical properties of the sensational world does not exist as a sentience cannot possess an inherent control over the sensational world from which its sentience must be derived and the concept of being both infinitely self-sufficient yet insufficiently enabled to naturally prohibit conditions requiring the need for intervention involve mutually exclusive conditions to prohibit their simultaneous coexistence.

11:7

Any imagined immortal beings with any inherent control over any physical properties of the sensational world do not exist as infinite beings and finite beings with control over properties to which they themselves are inherently subject to cannot exist.

11:8

Imaginary immortal inhabitants of either the mountain tops or the sky with inherent control over the physical properties of the sensational world referred to as gods are not publicly known to have any possibility of existence and cannot be honestly asserted to exist outside of delusion or of actual public demonstration of their existence involving observable independent responses to the observable world from which their independent presence can be deduced.

11:9

An individual's identification of a core feeling with actual knowledge pertaining to the presence of a claimed god is a delusional identification that fails to distinguish the possession of actual knowledge from private psychological experiences and corresponds to the conception of a subconscious superego.

11:10

The psychological experience of feeling the presence of a god that cannot be publicly demonstrated to consistently coincide with observable independent reactions of the observable world cannot be publicly known to be anything more than a psychological manifestation of the individual claiming to have the experience.

11:11

A god confined to an individual's choice of identifying a strong feeling with the presence of that god is a god whose existence is confined to the private psychological experiences of that individual's choice identifications.

11:12

A god confined to communicating messages through the private psychological experiences of an individual rather than directly to all of the individuals for whom the messages are alleged to be intended is a god whose existence is confined to the private psychological experiences of the individual communicating the messages.

11:13

A god confined to appearing privately to an individual rather than publicly to all for whom the individual alleges that the god desires to be known by is a god whose existence is confined to the private psychological experiences of the individual to whom the god appears.

11:14

A god confined to an individual's ascription of an unobservable sentience pertaining to an observable state of affairs is a god whose existence is confined to the private psychological experience of that individual.

11:15

A god whose proof of existence is confined to postmortem appearances in ancient, anonymous, non-eyewitness, accounts written two generations after the alleged appearances by individuals alien to the culture of the alleged god is a god whose existence is confined to the literary natures of such accounts.

11:16

The nature of an individual's god that fails to bring about declared future events is the nature of an undependable god that is neither omnipotent nor omniscient being confined to the individual's personal abilities of reason and knowledge of events.

11:17

A god confined to the private psychological experiences of an individual is a god whose existence is confined to the imagination of that individual.

11:18

A god whose existence is predicated upon non-existent states of affairs is a non-existent god.

11:19

The concept of a god is essentially the concept of a responsive sentience with inherent control over an aspect of nature that has the power to grant favors and has in some way interacted with the observable world thereby revealing the notion of its existence.

11:20

Any real state of affairs that cannot be publicly demonstrated to be a responsive sentience cannot be publicly known to be a god.

11:21

Anything sentient of the observable world is itself observable as it must possess a neural network tangible to the physical properties of the sensations received from the observable world.

11:22

An omnipresent god, sentient of the observable world, must possess an omnipresent neural network tangible to the physical properties of the sensations received from the observable world thereby making the omnipresent neural network an observable neural network and the omnipresent god a universally observable god.

11:23

A universally observable god that cannot be observed cannot exist by virtue of the mutually exclusive states of affairs to define it.

11:24

Gods that cannot be observed in the observable world cannot be real in the observable world.

11:25

Gods that can only be observed in private psychological experiences or only be observed in ancient imaginary tales are gods that can only be imaginary gods.

Dishonesty

12:1

Unsubstantiated assertions of either validity or possibility for imaginary states of affairs that an individual desires to be real are the basis for the expression of ideological dishonesty.

12:2

No state of affairs can be known to be real beyond possession of an actual knowledge of the state of affairs as to know a state of affairs is to have an actual knowledge of the state of affairs.

12:3

The only real state of affairs a person can know and confirm that no other person can know and confirm is their own psychological experience.

12:4

Any real state of affairs outside of a person's psychological experience that can be known and confirmed by that person can be shown to and confirmed by any other person.

12:5

Phenomenal consistency is the basis upon which the real states of affairs of the observable world are distinguishable from the inconsistent states of affairs of private psychological experience (e.g. dreaming and hallucination) and upon which the real states of affairs of the observable world are distinguishable from the imaginary states of affairs produced by either dishonesty or delusion.

12:6

A claim of phenomenal validation upon the insufficient occurrence of a state of affairs known to have failed public demonstration of phenomenal consistency is a consciously dishonest claim.

12:7

An unsubstantiated ascription of a necessary association between two or more states of affairs is either a dishonest ascription or a delusional ascription.

12:8

Failure to distinguish one's belief from one's actual knowledge is a failure to distinguish an imagined state of affairs from a real state of affairs.

12:9

Claimed belief in a state of affairs that the claimant recognizes to be contrary to states of affairs already recognized by the claimant as being true is either a false claim to belief or a delusional belief.

12:10

Any state of affairs asserted to be real outside of a person's private psychological experience and beyond any possibility of ever being publicly known is either a dishonest assertion or a delusional assertion.

12:11

Any state of affairs outside of a person's private psychological experience asserted to be real beyond any need for confirmation is either a dishonest assertion or a delusional assertion.

12:12

Feelings are not a means of determining real states of affairs outside of a person's private psychological experience.

12:13

A strong desire or feeling for a state of affairs being real outside of a person's private psychological experience is not knowledge of the state of affairs being real.

12:14

An unsubstantiated claim for an inherent ability to know or feel the truth of a state of affairs outside of a person's private psychological experience is either a dishonest claim or a delusional claim.

12:15

A substantiated claim for an inherent ability to know or feel the truth of a state of affairs outside of a person's private psychological experience is confirmed upon the phenomenal consistency of the publicly demonstrated ability thereby ruling out occurrences of statistical probability.

12:16

An unsubstantiated claim for a phenomenon in which states of affairs beyond one's control are brought into compliance with one's requests is either a dishonest claim or a delusional claim.

12:17

A substantiated claim for a phenomenon in which states of affairs beyond one's control are brought into compliance with one's requests is confirmed upon the phenomenal consistency of the publicly demonstrated phenomenon thereby ruling out occurrences of statistical probability.

12:18

An implication that an unsubstantiated answer is automatically validated by the absence of any other answers being immediately provided is a dishonest implication.

12:19

Merely possessing a desire for the possibility of a state of affairs is not an engagement of reason for the possibility of the state of affairs.

12:20

Misrepresenting the mere possession of desire for the possibility of a state of affairs as being an engagement of reason for the possibility of the state of affairs is either a sincere temporary confusion of the distinction between mere desire and actual reasoning or a dishonest attempt at representing an impossible state of affairs as being possible.

12:21

A claim to honesty based upon a sincere belief in an asserted state of affairs is an invalid claim when the state of affairs is asserted as being either real or possible without any knowledge of the state of affairs being real or possible.

12:22

Believing a state of affairs to be real while consciously not knowing the state of affairs to be real yet asserting the state of affairs to be real is dishonest.

12:23

Believing a state of affairs to be possible while consciously not knowing the state of affairs to be possible yet asserting the state of affairs to be possible is dishonest.

12:24

Any affirmation of a state of affairs that the affirmant sincerely believes yet consciously does not know to be affirmed is a dishonest affirmation.

12:25

To provide demonstration of the existence or possibility of a state of affairs publicly asserted to be real or to be possible is to respond honestly to the asserted state of affairs.

12:26

To consciously avoid demonstration of the existence or possibility of a state of affairs publicly asserted to be real or to be possible is to respond dishonestly to the asserted state of affairs.

12:27

Ignorant advocation of the sincere belief in a false state of affairs becomes a dishonest advocation upon the continued advocation of the false state of affairs subsequent to the believer becoming consciously aware of the falsity of the state of affairs.

12:28

Sincere attempts at validation of unsubstantiated assertions upon challenge to the assertions accompany sincere beliefs in the reality of the unsubstantiated assertions.

12:29

Conscious avoidance of opportunities to validate challenged unsubstantiated assertions implies conscious awareness of the assertions not being substantiated and dishonest intent behind the assertions.

12:30

Public assertions of private experiences that have convinced the asserter of extraordinary states of affairs remain publicly unsubstantiated and beyond sufficient consideration for public validation without public demonstration.

12:31

Assertions of public validation for extraordinary states of affairs that are only asserted to have been privately validated are dishonest attempts at having extraordinary claims publicly accepted without the sufficient reason or public demonstration the claimant knows to be required for public validation.

*

12:32

A non-historical character of imaginary attributes known

ultimately from imaginary tales is an imaginary character.

12:33

The possibility for the objective existence of one imaginary character can be no greater than the possibility for the objective existence of any other imaginary character.

12:34

An arbitrary identification of an imaginary sentience with a fundamental state of affairs without observable independent reactions from the fundamental state of affairs to identify the existence of the sentience is both a dishonest identification and an identification that is no more credible being asserted for one imaginary sentience than it is for being asserted of any other imaginary sentience.

12:35

A public advocation for the existence of an entity that cannot be publicly known is either a dishonest advocation or a delusional advocation.

12:36

Assertions for the objective existence of entities only known to be imaginary are either dishonest assertions or delusional assertions.

12:37

The fundamental grounds for dismissing the existence of any imaginary entity are the fundamental grounds for dismissing the existence of all imaginary entities.

12:38

The advocation for the existence of any imaginary entity to the exclusion of all other imaginary entities is either a dishonest advocation or a delusional advocation.

12:39

Knowing that the mere preference for the existence of an imaginary entity is the sole motivation for advocating the existence of the imaginary entity while pretending there to be objective reason for advocating the existence of the imaginary entity is a dishonest advocation.

12:40

An assertion that an imaginary entity is responsible for the existence of an observable subject is either a dishonest assertion or a delusional assertion.

12:41

An ascription of a state of affairs portrayed as being extraordinary that is ascribed to an imaginary entity with no observable affiliation to the state of affairs is either a dishonest ascription or a delusional ascription.

12:42

Claimed identification of the presence of an imaginary entity based upon a private feeling is either a dishonest identification or a delusional identification that fails to distinguish one's private psychological experience from objective knowledge of a presence and its identity.

*

12:43

Preference for an imaginary concept, no matter how strong, can never make the imaginary concept real.

12:44

Belief in the reality of an imaginary concept predicated upon the strength of preference for the imaginary concept is a delusional belief.

12:45

An assertion for the existence of an imaginary concept known to be impossible is either a dishonest assertion or a delusional assertion.

12:46

A sincere insistence upon the existence of a subject that the insister knows to be impossible is the expression of a delusion.

12:47

A wager for the possibility of a subject that the bettor knows to be impossible and unconfirmable is neither an honest wager nor a legitimate wager.

12:48

Ignoring the facts pertaining to an impossible state of affairs will not allow the state of affairs to be possible.

*

12:49

Good news declared to be at hand for a concurrent generation that fails to be realized in that generation is false news.

12:50

Subsequent re-declaration of a given declaration exposed as being a falsehood is a dishonest commitment to portraying the exposed falsehood as being true.

12:51

An event that does not occur within its time-dependent defined limit is an event that has failed to occur and therefore cannot occur.

12:52

Promoting the expectation of an event beyond its time-dependent defined limit is either a dishonest promotion or a delusional promotion.

12:53

Ignoring the fact of a portrayed predicted event's failure in order to encourage the false expectation of its future occurrence is a dishonest encouragement.

12:54

The concept of heaven as a place of habitation, prior to 1878, had always been publicly conceived of as being a tangible 'pound-out' at a physical distance in the sky as the reference of 'heaven' literally refers to the sky and had never been publicly conceived of as being an intangible hyperspace habitation prior to the publishing of 'Transcendental Physics' by Johann K. Zöllner in 1878.

12:55

A place of habitation that has been shown not to exist as a habitation at its space-dependent defined location is a place of habitation that does not exist.

12:56

Promoting the expectation of either reception from or arrival at a place of habitation that has been shown not to exist at its space-dependent defined location is either a dishonest promotion or a delusional promotion.

*

12:57

Unsubstantiated assertions of a degree of poetic quality only being possible upon having been inspired from a transcendental source are either dishonest assertions or delusional assertions.

12:58

A work composed of artificial sequences and unhistorical details with no eye-witness testimonies is a work of fiction.

12:59

An unverifiable account featuring elements of fiction is an intentionally imaginary account that provides no sufficient basis for acceptance as an account of actual events.

12:60

A single specific occurrence of an event of a single time and a single location cannot occur at more than one time or at more than one location.

12:61

Reports of a single specific occurrence of an event occurring at more than one time or at more than one location involves false reporting.

12:62

Conflicting accounts that all clearly disagree on the specific time and location of a central event cannot all be true accounts of the central event.

12:63

Asserting conflicting accounts that all clearly disagree on the specific time and location of a central event to all be true accounts of the central event is either a dishonest assertion or a delusional assertion.

12:64

Advocating unquestionable validity for fictional accounts publicly shown to be anonymous, non-eyewitness, plagiarized, stories of fundamentally conflicting claims and fictional attributes, known to have originated generations after the claimed events from a culture and region alien to the culture and region of the claimed events is either a dishonest advocation or a delusional advocation.

12:65

Denials of the plain meanings of texts that are obviously erroneous though asserted to be infallible are dishonest denials exposing the denier's lack of genuine faith in the validity of the texts.

12:66

To consciously ignore or deny the plain meaning of a text in order to assert a desired meaning in accordance with a pre-conceived ideology that the text plainly does not support is to engage in an act of dishonesty.

12:67

The conscious promotion of fiction as reality to deceitfully influence the life decisions of others is a socially subversive endeavor of dishonesty.

*

12:68

Religion is the social devotion to an imaginary-based worldview of non-confirmable claims asserted to be true.

12:69

Any assertion of truth beyond any reason or observation of what can be known to be true is inherently dishonest.

12:70

Social devotions to worldviews of non-confirmable claims asserted to be true beyond any reasons or observations of what is known to be true are inherently dishonest.

12:71

Encouraging individuals to accept states of affairs known to be without any sufficient basis for acceptance is a conscious endeavor of dishonesty.

12:72

Encouraging individuals to accept overall states of affairs that one does not know to be true yet does know to involve false states of affairs is a conscious endeavor of deception.

12:73

An occupation that requires either lying or arguing in defense of falsehoods is a dishonest and immoral occupation.

12:74

The suspension of honesty for commitment to a belief is a dishonest and immoral commitment.

12:75

Any religion that functionally requires the suspension of honesty for commitment to a belief is a dishonest and immoral religion.

12:76

Ongoing social attempts at maintaining religious credibility in the modern world repeatedly involve religious adherents making concessions to the findings of science though the findings of science are never found to be concessions made to the extraordinary claims of religion.

12:77

The mere desire to believe in extreme improbabilities and the falsehoods of fiction to justify imagined postmortem states of affairs facilitates the social embrace of delusional perspectives in non-delusional individuals.

12:78

Non-delusional individuals who accept the facts of their own mortality may be personally more inclined towards pursuing ways of improving the overall quality and duration of life in general while non-delusional individuals who ignore the facts of their own mortality for a retreat into the delusion of a postmortem immortality may be personally less inclined towards pursuing ways of improving the overall quality and duration of life in general.

12:79

An individual's educational background and standard of living can have a social influence upon the degree to which they are likely to ignore the facts of their own mortality for a retreat into the delusion of a postmortem immortality.

12:80

Higher degrees of education and higher standards of living can facilitate greater degrees of attention to the facts of mortality and social pursuits of improved overall quality and duration of life.

12:81

Support of political policies corresponding to increased inaccessibility of higher degrees of education and higher standards of living have political correspondences with social support for economic and militaristic foreign impositions and the legal validation of preferred religious perspectives in domestic society.

12:82

Support of political policies corresponding to increased inaccessibility of higher degrees of education and higher standards of living that also correspond to selective denials of reason and physical evidence supporting ecological considerations for socioeconomic reform is support derived from conscious choices to ignore social responsibilities that impose restrictions upon personally preferred lifestyles.

12:83

The issue of whether or not to relate to the world on the basis of reason and observation is not an essential issue of belief and disbelief but is an essential issue of honesty and dishonesty.

12:84

Non-delusional individuals who do not acknowledge clearly revealed inconsistencies or contradictions of a specific matter are clearly revealed to be dishonest individuals in regards to the matter.

12:85

Personal decisions to ignore undesired objective information in order to perpetuate desired assertions of unreasonable claims are emotional, consciously dishonest, decisions that expose unbelief in the very claims for which the objective information is willfully being ignored.

12:86

Non-delusional individuals who advocate the reality of imaginary states of affairs in dismissal of reason and observation are individuals consciously engaged in endeavors of dishonesty.

12:87

Faith that must be maintained by a selective denial of the use of reason and of physical evidence that is otherwise commonly accepted in all other areas of life is faith that is either emotionally dishonest or clinically delusional.

12:88

Most individuals to make unsubstantiated assertions for either the validity or possibility of a desired state of affairs while denying reason and observations to the contrary are not clinically delusional.

12:89

Most individuals to make unsubstantiated assertions for either the validity or possibility of a desired state of affairs while denying reason and observations to the contrary are reacting emotionally with dishonesty to avoid facing undesired truths about the world that deprive them of their sense of security in believing to have known the world.

Section Two:

R-Theory

Gravitation

1:1

Acceptance of general relativity invalidates the pursuit of a quantum theory of gravitation.

1:2

The gravitation of general relativity is not an interaction between masses but is a phenomenon resulting from the contraction (i.e. bending) of space by the quantum force (i.e. the collective spectrum of quantum phenomena resulting in energy/mass) and is therefore the quantum-space interaction that has no independent existence from the overall fundamental existence of the quantum force thereby inherently rendering gravitation as being a quantum force.

1:3

All quantum forces convey the force of gravitation as a universal property of the quantum force as gravitation is intrinsic to each force and has no independent existence from them as gravitons.

1:4

Whereas quantum mechanics describes the quantum interactions *in* space, general relativity describes the overall nature of the very same quantum interactions *with* space (thereby necessitating the inclusion of space in the Standard Model if gravitation is to be accounted for).

1:5

As with the electric and magnetic aspects of the electromagnetic wave, nothing is necessary for reconciling the two aspects of the quantum interactions *with* and *in* space as they are dimensionally distinct in their domains of interaction and cannot be treated as being of the same type of interaction as has been pursued in proposed quantum theories of gravitation.

1:6

The quantization of mass hereby referred to as the quantum, resulting in values of both mass and radius equal to Gh/c^2 with a Compton wavelength of c/G, is the smallest possible measure of mass corresponding to the smallest possible force of gravitation.

Time

2:1

Time is the absolute value of all motion and is coextensive with space from the frame of reference that is within the universe as space is both permeated by energy and is itself undergoing the motion of expansion.

2:2

The dilation of relative time for an accelerating mass is merely the proportionally restricted motion of a contracted space.

2:3

As a feature of local motion, relative time is intrinsically identifiable with local quantum activity that is merely a subset of overall time and should therefore never be utilized as an unqualified reference that is easily mistaken for a reference to overall time.

2:4

Within the universe frame of reference, the arrow of time cannot reverse without a force capable of reversing the fundamental forces governing all motion from the quantum level to the cosmological expansion.

Space

3:1

As the properties of luminal mass and Schwarzschild mass are identical, just as no amount of mass can ever be accelerated to the velocity of light, no amount of mass can ever be compressed to its Schwarzschild radius implying that any supposed black hole is actually a kakesu (dark star) composed of an as yet unidentified state of matter.

3:2

$V \rightarrow c = G \rightarrow r_s.$ In regarding all acceleration as being indistinguishable, the Equivalence Principle requires an application of the Lorentz factor to gravitation (gamma G) in which V (velocity) becomes V_e (escape velocity) with the result that mass with a V_e of c can never be obtained without infinite mass just as mass with a V of c can never be obtained without infinite energy.

3:3

Neither luminal mass nor Schwarzschild mass can ever be obtained due to the compressive strength of space itself that resists being compressed to the luminal limit thereby revealing c to represent both the velocity of energy and the compressive strength of space.

3:4

The resistance of space and therefore mass to be compressed to the luminal limit implies both a repulsive force of space that prohibits a luminal degree of compression as well as a fermion binding force that holds the individual quantum masses in close proximity with one another to form the various masses of quantum particles.

Motion

4:1

Motion is change in space.

4:2

The speed of light is the limit to the speed of motion.

4:3

Mass compressed to the degree that there is no motion is mass compressed to the degree that there is no space for change to take place and therefore no space for the activity of anything confined to the speed of light.

4:4

'No space' not only means the elimination of all of the space of a mass but also means the elimination of all of the mass from space (making r_s the boundary between space and hyperspace).

4:5

In accordance with the Uncertainty Principle, the potential for change in space is required for the existence of mass in space as mass is change in space (i.e. energy) confined to a volume of space.

4:6

Any force capable of completely suspending all motion in a

volume of space and therefore the motion of luminal velocity is a force that necessarily eliminates that volume of suspended motion from space.

4:7

In accordance with the Conservation Principle, as energy/mass can neither be eliminated from space nor infinitely increased within space, equations to result in 0 space of ∞ degree properties either imply an impossible state of affairs that naturally can never exist or the existence of a state of affairs outside of space possessing infinite degree qualities.

4:8

The properties of any theoretical physical phenomenon found to be in violation of physical principles are logically invalid until the physical principles that the properties violate are found to be invalid or the properties of the theoretical physical phenomenon are found to be in conformity with the physical principles or the properties of the theoretical physical phenomenon are physically verified to exist.

Gravitational Singularities

5:1

A conceptual absurdity arises from the fact that gravitational effects within an event horizon cannot have continuity with gravitational effects outside of an event horizon since the speed of gravitation is the speed of light.

5:2

As the curvature of space produced by a gravitational singularity is limited to traveling out at the speed of light, such curvature is unable to escape the event horizon thereby producing a hole in space that cannot gravitationally affect anything outside of the event horizon.

5:3

Conceptually, mass contracted to its Schwarzschild radius becomes a 2-dimensional sphere as a result of its completely contracted length thereby never becoming a gravitational point singularity of zero volume and infinite density.

5:4

A 2-dimensional sphere of infinite density with no interior for anything to fall into and absolutely no affect upon space exterior to its surface where all 3-dimensional objects instantaneously become 2-dimensional and the complete dilation of relative time renders everything motionless making all quantum positions and momentum simultaneously certain in violation of the Uncertainty Principle is a conceptual absurdity.

5:5

An object of no interior mass and no exterior curvature of space is an object of no gravitation and therefore no event horizon and no presence in space.

5:6

An imagined occurrence of mass that produces an event horizon immediately results in the mass' ejection from space with an equal and opposite force against the exterior of space in contribution to the cosmological constant.

Cosmological Constant

6:1

The cosmological constant is not the product of the vacuum energy as the vacuum energy is a product of the overall quantum field and not space itself with which it can only interact gravitationally.

6:2

The source of the cosmological constant is external to the space it expands, not intrinsic to it as, like gravitation, the cosmological constant is an interaction *with* space, not *in* space.

6:3

An overwhelming space-expanding force from within space must automatically overwhelm the force of gravitation with which it shares the very same dimensions of space occupied by the source of gravitation and therefore have at least the same degree of interaction with energy/mass that gravitation has thereby immediately resulting in a dark, starless, universe of dispersing dark matter, hydrogen, helium, lithium, and beryllium.

6:4

The conclusion that the cosmological constant is the indirect result of an eternal inflation of infinite false vacuum undergoing sporadic decays into true vacuums requires that the infinite false vacuum be both eternal and non-eternal.

6:5

An infinite false vacuum must be eternal in order to perpetually facilitate the pre-existent conditions of true vacuums.

6:6

The occurrence of true vacuums prohibit the nature of an infinite false vacuum from being uniformly eternal.

6:7

An infinite false vacuum must either eternally remain a false vacuum or eventually become a completely true vacuum as there are no grounds for the nature of the false vacuum to be both eternal and subject to decay.

6:8

An eternal infinite false vacuum is in reality a true vacuum by virtue of never being able to decay into a true vacuum while a non-eternal false vacuum must have a point of origin thereby prohibiting it from being infinite or being the source of anything more than a single inflation.

6:9

Spontaneous false vacuum decay along with ongoing vacuum energy cannot account for the cause of the cosmological constant.

6:10

The bounded elastic nature of space apparent through gravitation and the cosmological constant implies the existence of an unbounded hyperspace from which such elasticity is sustained.

Hyperspace

7:1

An omnidirectional pulse wave of bounded space spreading out in an unbounded 3-dimensional hyperspace of inverse quantum forces models a description of the cosmological expansion.

7:2

The forces of gravitation and the expansion each affect space without direct cancellation of the other's opposing influence.

7:3

The force of the expansion appears to be dimensionally perpendicular to the force of gravitation in that the force of the expansion is overwhelmingly greater than the force of gravitation without resulting in the direct cancellation of gravitation's opposite influence.

7:4

An intra-space contracting force coexisting with an overwhelming exo-space expanding force with which it shares a temporal point of origin implies the production of an omnidirectional pulse wave of bounded space within an unbounded hyperspace in which the limited quantum force of gravitation derived from within space is the inversion of the unlimited force of expansion derived from outside space resulting in an expanding toroidal universe.

7:5

An inverse identification of gravitation derived from within space with an expansion derived from outside space implies that the properties of hyperspace are the inversions of the quantum forces within space further implying that the nature of the forces are relative with respect to their positions with space (i.e. space = $r > r_s$ while hyperspace = $r < r_s$).

7:6

The inclusion of space and hyperspace with the Standard Model is the Relativistic Model.

7:7

On the basis of the Relativistic Model, understanding the origin of the universe and the existence of a multiverse is dependent upon understanding the properties of hyperspace (i.e. the development of a Relativistic or R-theory of quantum forces), not in a theoretical pursuit of quantum gravitation.

Section Three:

Standard Radix

1:1

Where p is any prime number, $p > 3 = 6x \pm 1$.

1:2

The **Sunu** (a z formed with a concave top bar), represents the circumference of a circle divided by its radius and should therefore be used in place of π and especially 2π as radius, not diameter, is the more fundamental unit of geometric measure.

1:3

Given a set of regular polygons, 6 is both the maximum amount of least sided polygons (i.e. equilateral triangles) that can be arranged around a single point and the maximum amount of infinitely sided polygons (i.e. circles) that can be arranged around another.

1:4

As a fundamental unit of geometric quantities and the numerical unit for the location of prime numbers, 6 is a natural unit of division upon which a standard radix should be divisible.

1:5

The greatest number of factors increase with the even multiples of 6 thereby revealing units of 12 to be the most factorable quantities with the numerical implication of 12 as being the natural radix for standard use.

Quantum Base Units

1 **Sitay** = $Gh/c^2(12^{57})$ = 16.041 joule.

1 **Shi** = $Gh/c^2(12^{56})$ = 1.336 kilograms.

1 **Shiay** = $Gh/c^2(12^{56})$ = 1.336 meters.

1 **Showp** = $Gh/c^3(12^{63})$ = 0.159 seconds.

If quantum based measures are ever adopted for SI units and standard use, the showp would only be used for scientific measures. For standard use, base 12 divisions of the day would result in 12 hours of 120 minutes each, composed of 144 minutes of 50 seconds each, composed of 144 seconds of 0.3472 seconds each.

Mejasinu Numerals

To avoid confusion between the use of a base 12 system and a base 10 system, the utilization of the base 12 system should employ a different numeral system such as the **Mejasinu** numerals.

ᐤΙץ‍ץᘓᒹᒷᔒᔓᒋᔭᐟᐠ

0 1 2 3 4 5 6 7 8 9 10 11

Although the names of the numerals would remain the same in every language, the Mejasinu names for the numerals pertaining to the quantities of ten and eleven would consistently remain **meja** and **mejawa** respectively for all languages.

www.ingramcontent.com/pod-product-compliance
Lightning Source LLC
Chambersburg PA
CBHW060412190526
45169CB00002B/872